The Universe is Not

Dying

i

The Universe is Not

Dying

A unified physics theory explaining the mysteries of dimensions, space, strings, matter, energy, light, time, particle spin, wave formation, black holes, quasars, and the energy-matter cycle

James L. Jordan, PhD, PhD & Deovina N. Jordan, PhD, MD

Copyright Page

Dedication

We dedicate this book to:

All Who Seek A Better Understanding Of The World and
Universe Around Them

James L. Jordan and Deovina N. Jordan

Prologue

This book about the universe discusses important topics in physics such as space, strings, particle spin, wave formation, black holes, quasars, and the energy-matter cycle. It is intended for:

- scientists (primarily physicists, astrophysicists, astronomers, mathematicians, statisticians, and other scientists in related disciplines),
- philosophers,
- professors,
- engineers, and
- students of higher learning.

It is also intended for people across nations and cultures - who seek a better understanding of momentous topics in physics such as time, dimensions, dual nature of light, and the origin and fate of the universe.

We would like to warn readers that this book is not entertaining, in terms of personal experiences and anecdotes. As such, it is not structured like the large number of books available in the bookstores today which are full of entertaining, amusing stories, but with little true substance and material, particularly for people who want to learn more about critical issues in physics and the related sciences (e.g., astronomy, biology, chemistry, geology). Rather, the texts of this book are written along the lines of scientific books and have more substantive content on the aforementioned topics. Hence, the texts are highly educational and very informative.

This book on the universe is intended to be thought-provoking to challenge people to think about hypotheses and theories, most of which are currently being presented as facts (e.g., the nature of time and the one-particle big bang theory). This will help increase intelligence as psychologists state that the more dimensions one can visualize, the more intelligent one becomes. Hopefully, that will be one reward to those who see the world around them in more than the popularly presented and commonly utilized four dimensional model of the space-time continuum.

Overview

Introduction

Is the universe really dying? One scientific theory proposes that the universe will die when maximum entropy is reached. In other words, the universe will end in a big freeze (a heat freeze), the opposite of the big bang. As such, it may even undergo a big collapse. Such is dependent upon the decaying of protons and does not entertain the renewal of energy and the recycling of energy and/ or matter. This all makes sense when considering the laws of thermodynamics along with the Theory of the Big Bang origin of the universe. In fact, there are many manuscripts penned about this topic - end of the universe. Yet, there aren't any books written about the universe not dying. This book is the first to discuss the concepts of why the universe is not in decline (or dying).

Book Description

This book presents a unique and innovative theoretical

approach to a number of important topics in physics and astronomy regarding the universe, namely: the dimensions, space, strings, matter, energy, light, time, particle spin, wave formation, black holes, quasars, and the energy-matter cycle.

This book:
- Explains how more than four dimensions can develop and be visualized.
- Elucidates how interactions of strings give rise to matter and anti-matter.
- Clarifies why time cannot be passed off as the fourth dimension.
- Explicates that time is associated with the ratio of energy and matter (Discussed in the ***Jordans' Law of Time***).
- Enlightens readers on the space-time continuum.
- Examines dark matter, dark energy and light.
- Describes how waves form from spinning particles (Discussed in the ***Jordans' Three Laws of Particle Spin and Wave Formation***).
- Reviews the significance of black holes and quasars in the origin and fate of the universe.
- Addresses how energy and matter are recycled in the universe in an energy-matter cycle (Discussed in the ***Jordans' Energy-Matter Cycle***).

It is erroneous to conceive space-time as four dimensions. The space-time continuum is more complicated than the four dimensions would indicate. The book explains the development of dimensions (the 0 dimension, the positive dimensions and the negative dimensions). It describes that there are strings which, upon interacting with other strings, are left without any, or with reduced, points of interaction or POI. This is the property that gives rise to matter and anti-matter (crystallized forms of energy). The book clarifies why time is not the fourth dimension as time is actually linked to interactions among dimensions. It explicates that time is associated with the ratio of energy and matter as explained in the *Jordans' Law of Time*. It also discusses the following: (1) Time is affected by factors such as gravity and temperature; (2) Dark matter consists of components of light matter at absolute zero (at which atomic motion ceases to exist) and has no time component in itself; (3) Dark energy is negative energy associated with negative dimensions; and (4) Light/ electromagnetic radiation possesses properties of both particulate and wave nature. The book describes how waves form from spinning particles with resulting differences in wavelength and wave amplitude in the *Jordans' Three Laws of Particle Spin and Wave Formation*. It reviews the significance of black holes and quasars. Black holes are responsible for the origin of the universe as well as its continued existence. They function as recycling centers for the galaxy. Quasars result when energy and matter escape the black holes. They function as the actual recycling processes in action. Finally, the book addresses how energy and matter are reprocessed in the universe in accordance with the *Jordans' Energy-Matter Cycle*.

The concepts discussed in this book indicate that the universe is not in decline (or dying). It will last far longer than often reported.

Table of Contents

Chapter One: The Theory of Dimensionality:

A Unified Theory of Dimensions, Space, Strings, Matter,

Energy, Light, and Time

Part 1: The Dimensions

Copyrights © 2011, 2019, 2020, James L. Jordan, PhD, PhD

and Deovina N. Jordan, PhD, MD

The Collapse of the Dimensions: One view of the dimensions is that there are three basic dimensions of space and a fourth of time. Basically, whatever happens in a dimension, such as the movement of three dimensional objects in three dimensional space, remains in that (three) dimensional phenomenon. If a two dimensional plane interacts with a three dimensional space, the highest dimension is called upon and the phenomenon becomes

1

three dimensional. But, if no interaction occurs, time would cease to exist. Time is not a dimension in and of itself; time represents the degree to which interactions occur within and among the three dimensions. Moreover, the observed time is still composed of the elements of actual time which occur from the interactions within and among those three dimensions. Time remains a relative concept since the interactions within and among those dimensions are not necessarily constant but can be changed/ influenced (e.g., by gravity). While the collapse of the dimensions into three may be one way of viewing them, it has the unfortunate effect of discarding much information of what is occurring within and among dimensions. Furthermore, it leads to the concept of time not being really defined, but allocated to some dimension in and of itself.

The Positive Dimensions (or the Dimensions): It is erroneous to conceive space-time as four dimensions. The space-time continuum is more complicated than the four dimensions would indicate. The first thirteen (0 to 12) dimensions can be conceived as follows (although 0 is not a positive dimension, or even an anti-dimension, it is included in this discussion to provide a starting point):

0 dimension – has no measurable attributes. This dimension would consist of pure energy and is even more basic than strings, for strings are really points of energy. Basically, it is an infinitesimally small point. The closest thing to a point is a string.

1st dimension – This is the line. It can be represented by a vector. It can transpire as movement occurs in a single direction without any variance whatsoever.

2nd dimension – This is the plane. It can be a collection of lines. It can also arise from the interaction of two lines.

3rd dimension – This is space with length, width and depth. It can be unmoving or it can arise from the interaction of a line with a plane.

4th dimension – This is the result of the interaction of two planes (both 2nd dimensions) or the interaction of a line (1st dimension) with the 3rd dimension. An example is a line moving within a three dimensional space.

5th dimension – This is the result of the interaction of the 2nd dimension with the 3rd dimension. An example would be a plane (the 2nd dimension) moving within space (the 3rd dimension).

6th dimension – This is the result of the interaction of the 3rd dimension with the 3rd dimension. An example would be an object having three dimensional properties in which the points therein (the 0 dimension) are also moving about in a space occupying length, width and depth.

7th dimension – This is the result of the interaction of the 6th dimension with the 1st dimension. In other words, it becomes the rate, or velocity, at which the two 3rd dimension components interact and move simultaneously in a single direction. But, that is in only one single linear component. It is here where time actually enters the dimensional picture.

8th dimension – This is the result of the interaction of the 6th dimension with the 2nd dimension. In other words, it becomes the

3

rate, or velocity, at which the two 3rd dimension components interact and move simultaneously in two directions. But, that is within a planar component. This is also a dimension in which time plays a role.

9th dimension – This is the result of the interaction of the 6th dimension with the 3rd dimension. In other words, it becomes the rate, or velocity, at which the two 3rd dimension components interact and move simultaneously in three directions. But, that is within a three dimensional component. The 9th dimension also arises from the change of the velocity, adding another 1st dimensional component, as it interacts with the 8th dimension. This is also a dimension in which time plays a role.

10th dimension – This is the result of the interaction of the 9th dimension with a 1st dimension factor. In other words, it includes the change in the rate, or acceleration/ deceleration, within the other dimensions, as it pertains to points moving in a single direction without any variance whatsoever. This is also a dimension in which time plays a role.

11th dimension – This is the result of the interaction of the 9th dimension with a 2nd dimension factor. In other words, it includes the change in the rate, or acceleration/ deceleration, within the other dimensions, as it pertains to points moving in a plane or two directions. This is also a dimension in which time plays a role.

12th dimension – This is the result of the interaction of the 9th dimension with a 3rd dimension factor. In other words, it includes the change in the rate, or acceleration/ deceleration, within the other dimensions, as it pertains to points moving in three directions. This is also a dimension in which time plays a role.

The Negative Dimensions (or Anti-Dimensions) and Anti-Matter: It is erroneous to conceive space-time as four dimensions. The space-time continuum is more complicated than the four dimensions would indicate. The dimensions also include those which, when directly interacting with other dimensions, cancel out, in part or wholly, the characteristics of the other dimension(s). The first thirteen anti-dimensions (0 to12) can be conceived as follows (although 0 is not an anti-dimension, or even a positive dimension, it is included in this discussion to provide a starting point):

<u>0 dimension</u> – having no measurable attributes. This dimension would consist of pure energy and is even more basic than strings, for strings are really points of energy. Basically, it is an infinitesimally small point. The closest thing to a point is a string.

The 0 dimension has mirror images, positive and negative, along with a completely neutral form (in which the strings have no point of interaction or POI). There are strings which, upon interacting with other strings, leave both strings without any, or with reduced, POI. It is this property that gives rise to matter and anti-matter. Neither matter nor anti-matter cancel themselves out. But, when the two interact, their dimensional properties cancel out. That is because the strings in the matter interact, in a canceling manner, with the strings in the anti-matter, eliminating the POI in both the positive strings (which form matter) and the negative strings (which form anti-matter).

1st anti-dimension – This is the line. The properties appear to be identical to the 1st dimension. It can be represented by a vector. It can occur as movement occurs in a single direction without any variance whatsoever. In contrast to the 1st dimension, the 1st anti-dimension would cancel out, when interacting with, the 1st dimension. If a 2nd dimension interacts with a 1st anti-dimension, a 1st dimension results. If a 1st dimension interacts with a 2nd anti-dimension, a 1st anti-dimension results.

2nd anti-dimension – This is the plane. It can be a collection of lines. The properties appear to be identical to the 2nd dimension. It can also arise from the interaction of two lines. In contrast to the 2nd dimension, the 2nd anti-dimension would cancel out, when interacting with, the 2nd dimension. As mentioned earlier, if a 1st dimension interacts with a 2nd anti-dimension, a 1st anti-dimension results. But, if a 2nd dimension interacts with a 2nd anti-dimension, then 0 dimension results. The interactions between dimensions and anti- dimensions follow the same manner as the addition of positive and negative numbers. As such, which dimension results from which interaction will not be discussed in detail herein.

3rd anti-dimension – This is space with length, width and depth. The properties appear to be identical to the 3rd dimension. It can be unmoving or it can arise from the interaction of a line with a plane.

4th anti-dimension – This is the result of the interaction of two planes (both 2nd anti-dimensions) or the interaction of a line (1st anti-dimension) with the 3rd anti- dimension. An example is a line moving within a three dimensional space.

5th anti-dimension – This is the result of the interaction of the 2nd anti-dimension with the 3rd anti-dimension. An example would be a plane (the 2nd anti-dimension) moving within a three dimensional space (the 3rd anti-dimension).

6th anti-dimension – This is the result of the interaction of the 3rd anti-dimension with another 3rd anti-dimension. An example would be an object having three dimensional properties in which the points therein (the 0 dimension) are also moving about in a space occupying length, width and depth.

7th anti-dimension – This is the result of the interaction of the 6th anti-dimension with the 1st anti-dimension. In other words, it becomes the rate, or velocity, at which the two 3rd anti-dimension components interact and move simultaneously in a single direction. But, that is in only one single linear component. It is here where time actually enters the anti-dimensional picture.

8th anti-dimension – This is the result of the interaction of the 6th anti-dimension with the 2nd anti-dimension. In other words, it becomes the rate, or velocity, at which the two 3rd anti-dimension components interact and move simultaneously in two directions. But, that is within a planar component. This is also an anti-dimension in which time plays a role.

9th anti-dimension – This is the result of the interaction of the 6th anti-dimension with the 3rd anti-dimension. In other words, it becomes the rate, or velocity, at which the two 3rd anti-dimension components interact and move simultaneously in three directions. But, that is within a three dimensional component. The 9th anti-dimension also arises from the change of velocity,

7

adding another 1st anti-dimensional component, as it interacts with the 8th anti-dimension. This is also an anti-dimension in which time plays a role.

10th anti-dimension – This is the result of the interaction of the 9th anti-dimension with a 1st anti-dimension factor. In other words, it includes the change in the rate, or acceleration/ deceleration, within the other dimensions, as it pertains to points moving in a single direction without any variance whatsoever. This is also an anti-dimension in which time plays a role.

11th anti-dimension – This is the result of the interaction of the 9th anti-dimension with a 2nd anti-dimension factor. In other words, it includes the change in the rate, or acceleration/ deceleration, within the other dimensions, as it pertains to points moving in a plane or two directions. This is also an anti-dimension in which time plays a role.

12th anti-dimension – This is the result of the interaction of the 9th anti-dimension with a 3rd anti-dimension factor. In other words, it includes the change in the rate, or acceleration/ deceleration, within the other dimensions, as it pertains to points moving in three directions. This is also an anti-dimension in which time plays a role.

Dimensions Greater than the 12th - As noted earlier, higher dimensions arise from the interactions of lower dimensions. Also noted, time reflects the degree of the interaction between or among dimensions. As such, dimensions can arise much higher than the 12th. For the sake of simplicity, only interactions between 3rd dimensions will be discussed in this section. Take for

8

example the universe. It is expanding. As such, it exhibits the interaction of a 3rd dimension with another 3rd dimension. This results in the 6th dimension. But, the galaxies are moving in the universe, resulting in another 6th dimension. The combined effect is the 12th dimension. But, within the galaxy, components move. This is another 6th dimension. Ergo, the result is the 18th dimension. This has all happened even before addressing what is happening on the surface of a planet or within a living body. One result is that time is layered, becomes fractionalized into components, but is only observed as the summation of the different components of actual time. In terms of observational time, components of actual time appear constant; these include those components dealing with expansion of the universe, movement of the galaxy in the universe, and movement within the galaxy. This is particularly evident since humanity is actually viewing the interaction from an extremely narrow duration (or part of the interaction among the dimensions) and limited space. Thus, the 12th dimension, as we observe it, can be thought of in terms of the definition given earlier.

Chapter One: The Theory of Dimensionality:

A Unified Theory of Dimensions, Space, Strings, Matter,

Energy, Light, and Time

Part 2: Strings, Matter and Anti-Matter

Strings, Matter, and Anti-Matter: There are strings which, upon interacting with other strings, are left without any, or with reduced, points of interaction or POI. The same can be said about the other strings which they have interacted with (discussed in greater detail later). It is this property that gives rise to matter and anti-matter; both matter and anti-matter can be thought as being crystallized forms of energy (or uniform packets of energy), albeit different. When matter and anti-matter interact, neither matter nor anti-matter directly cancel themselves out. The result is an indirect effect from the interaction of their dimensional properties. When matter and anti-matter interact, their dimensional properties are the components that cancel out. That is because the strings in the matter interact, in a canceling manner, with the strings in the anti-matter, eliminating the POI in both the positive strings (which form matter) and the negative strings (which form anti-matter). As such, what we perceive as matter has one kind of string ("positive string/s"); but, if a string is anti-matter (a mirror image of the positive string/s), it would be perceived as being anti-matter (having "negative string/s"). This would be much like superimposing a photographic image with the negative; when they are imposed on each other, the image is lost with the result being black. The strings of the matter and the anti-matter would cause negation of the POI, causing the dimensional state to be reduced to the 0 dimension (if all POIs are so negated). The release of energy from such a reaction results from the induction of POI on surrounding strings as the POI in the matter and anti-matter are neutralized.

Strings: A number of presentations of the strings in string theory indicate that the strings are more or less the same. If that were true, it would not be logical to transform strings, of about the same nature, into constant entities such as photons, electrons, protons, etc. In fact, the particles made from the strings have constant natures, such as mass. But, such constant natures cannot arise from random gathering of strings into clumps. There is order in what arises from the strings; there is order in the strings. That order arises from the strings having POIs. The POI serves as means by which strings cluster and form matter. It can be thought of as though the strings are pieces in a jigsaw puzzle. As for the

POI, they can be of odd number, even number, or even nonexistent on a string. Moreover, if present, they can have a relatively unstable nature, a relatively stable nature, a strong nature, a weak nature, or a combination of strong (at one part) and weak nature (at another part). The strings with strong, relatively stable POIs form the foundation for particles. But, they are surrounded by strings with POIs that have one region with a strong nature (to interact with the strong POI strings) with a weak region exposed to the external environment. The weak region, however, can be transformed into an unstable strong POI region interacting with the environment. The environment of the particles (e.g., electrons, neutrons, protons, etc.) consists of strings which are without POI. However, when particles interacting with the POIs on a string with POI, a secondary POI can be induced in such a string. By that means, a temporary link can be formed between the electron and the proton, causing them to stay in cohort with each other. However, that interaction is extremely brief and new interactions are formed. As such, the space between particles acts in a similar manner as a nonpolar solvent, but the particles act as polarized particles. The interaction of strings of different types accounts for the wide variety of interactions indicated in quantum mechanics. A notable interaction is that of the electron as it travels about the nuclei. Without the interaction of strings of different types, the path, particularly for a hydrogen atom, could be represented as being constant, similar to that of a planet around the sun. But, the path is not elliptical, or circular, but better represented by a cloud in which the exact path of the electron cannot be determined with any precision.

When chemical bonding occurs, it is the result of the

interaction of strings of different types that are stronger for one atom than for another. When the interaction is stronger in one atom than in another, then the induced POI can be stronger in that atom, affecting the electrons in another atom with a weaker overall induced POI. When the induced POIs do not extend much beyond the atom, then the atoms may resist forming chemical bonds (such as with the noble gases); in its most extreme form, the bonding becomes ionic. When the POIs are induced beyond the atom, but are basically equal among the atoms, then the resulting bonds are either covalent or metallic, with the metallic having the most equal POI induction between the atoms.

Within the nuclei of more than one proton, the POIs of the strings of adjacent protons and neutrons have a more stable relationship than that between the nuclei and the surrounding electron(s). The POIs on the surfaces are more matching in alignment, though not necessarily universally so. That alignment plays a crucial role in nuclear stability: the more matching in alignment, the stronger the bonding among particles within the nuclei. Within the protons and neutrons of the nuclei, the POIs do not always align perfectly either. This can cause a tendency of POIs to shift to seek better alignment; but that shifting is ongoing and may end up with the formation of weak, but transient POI, or a rather permanent mismatch (as between odd versus even POI in the strings, or between strings with a variance in the number of POI, though both may be even or odd numbered). The instability of the nucleus can also arise when strings with POI have both strong and weak regions or when strings with an unstable nature have shifting POI. This may result in the development of radioactivity in which a more stable relationship between the strings is sought.

The basically neutral strings (those without POI) serve another purpose than the role that they play between a nucleus and the surrounding electrons. The pulse of gravity can also be transmitted through them through the sequential development of weak POIs. This causes the warping in space, noted in Einstein's Theory of Relativity, as the pulse bypasses a mass which would have strings of strong POI. The interaction goes further as the gravitational pulse interacts with the distortion caused by the large collection of strings having that strong POI. This would occur at the cosmic scale since at the atomic and subatomic level, the interaction between the gravitational pulse would be too weak to be noticeable. It is the role of POIs, contained in the subatomic and atomic levels, to cause transient changes in strings without POIs, allowing them to interact.

Being in Two or More Places at the Same Time: It is possible for something (e.g., a particle) to appear to be in two or more places at exactly the same time. This is because a residual effect may be noted in the section on strings discussed previously. Because a signature may exist where something was, but appearing that it is still there, that something can end up appearing to be in two different places at one time, but, in reality, be in only one place at that particular moment in time. An analogy would be a footprint, which is the signature that the foot was or is at a particular location. But, the footprint could be mistakenly used as evidence that the foot is at that particular location. This may be true in some cases, but false in others.

Chapter One: The Theory of Dimensionality:

A Unified Theory of Dimensions, Space, Strings, Matter, Energy, Light, and Time

Part 3: Time and Space-Time Continuum

The Nature of Time: As noted earlier, time plays a role particularly in dimensions seven and higher. To begin with, time is not a constant. It only appears that way. But, time is actually the result of the interaction between matter in the dimensions and energy. Time, as it is commonly perceived and measured, is actually a measure of the rate by which observable dimensions interact. When matter and energy are relatively constant, the interactions between the dimensions remain fairly constant and, as such, time remains relatively constant. When only matter exists, then there are no such interactions. At that point, the rate of time is zero. But, that does not happen; when matter becomes too abundant, then it begins to reconvert back to energy. When there is virtually no matter, as at the onset of the big bang, then there is an overabundance of energy to facilitate interaction between the dimensions. At that point, the rate of time is virtually infinite. However, this too comes to an end as the amount of matter increases. Therefore, as the amount of energy increases relative to the amount of matter, the rate of time increases. Conversely, as the amount of energy decreases relative to the amount of matter, the rate of time decreases. To put these ideas in the form of an equation, the rate of time is: *t (Time) = e (Energy) ÷ m (Matter)*. This is the ***Jordans' Law of Time***. From the standpoint of the equation of $e = mc^2$, divide both sides by mc^2. Then, $1 = e \div mc^2$. The c^2 is the square of a distance of light traveled in a particular unit of time. The distance is a constant and can, for illustration purposes, be dropped from the equation; moreover, the c^2 portion can be substituted by time (t). The equation becomes $1 = e \div mt$. As such, the time component can be multiplied for both sides, resulting in the equation: *t (Time) = e (Energy) ÷ m (Matter)*. This is the ***Jordans' Law of Time***. Note: The equation demonstrates the nature of the relationship between energy and matter and how they pertain to time. Strictly speaking, the equation indicates that the rate of time is determined by the interaction of energy and matter (which influence the amount of interaction among the dimensions). The equation does not represent a strict mathematical equation that can be used to precisely determine the rate of time or the degree of interactions among dimensions. The reason for that is there are many dimensional interactions occurring simultaneously.

At the beginning of the universe, the universe expanded at a very high rate of time, since the preponderance of that existence was energy. That is why matter can be five billion years old in a universe which is fifteen billion years old. The expansion was fastest at the beginning of the universe and has slowed down since then. But, there are places in the universe in which time is slower and where time is faster than it is here. That is because the interaction between dimensions is not constant throughout the universe.

The Effect of Gravity on Time: According to the theory of relativity, gravitational forces on a celestial body (e.g., the earth) causes time to be slower than in the absence of gravitational forces (e.g., when in orbit around the earth). This was assumed to be a direct effect of gravity on time (which is used to be considered the fourth dimension of the space-time continuum). In actuality, gravity on the surface of a planet (or other body with sufficient gravity) restricts the interaction of some dimensions, namely those which would correspond to "up and down" directional interactions. The surface of the planet (or other body with sufficient gravity) simply stops (or sufficiently curtails) those "up and down" directional interactions; if not, one would not have something/ someone (e.g., a person) standing upon the surface of the planet earth. The reduction in dimensional interactions causes the slowing of time by gravity; as such, the impact of gravity is indirect. When something/ someone (e.g., a person) leaves the surface, the interactions (of what would be "up and down" relative to the earth's surface) resume and, consequently, those interactions become part of the overall component which, in summation, is the observed time.

But, does gravity always slow down time? No. In fact, when something approaches a black hole, it accelerates toward the black hole and into it. With lateral dimensional interactions (interactions lateral to the direction of the black hole), the dimensional interactions increase in the dimensions oriented parallel to the black hole. The net result is an increase in observed time (from the vantage point of the item involved, not from the vantage point of someone or something outside of the event horizon). But, when that object actually reaches the center of the black hole, the dimensional interactions are even more constrained than that at the surface of a planet (or other solar body with sufficient gravity). Observed time in the center of a black hole becomes very slow. The matter of the black hole is addressed later in this book.

The Effect of Temperature on Time: Gravity is not the only force that can alter the rate of time (by affecting the interaction of dimensions). Temperature, at extreme elevated intensities, would result in greater interactions (e.g., among atoms, particles). This would cause time to be faster. Conversely, time, at absolute zero, would be slower since the atomic and particulate interactions would decrease, resulting in a decreased interaction of dimensions.

Space-Time Continuum: The space-time continuum is actually the interaction of space (dimensions thereof); as such, time is associated with space. But, the time component is not a dimension in and of itself, but reflective of the interaction between the dimensions that comprise space. As such, time and space appear to be independent, but that is not true. Because space is not static, it appears to have the component of time.

Apparent time is what is actually observed. Apparent time is the summation of the interaction of all the dimensions; thus, it is the summation of all the subsets of time, since time is the degree to which two or more dimensions interact with each other. A component of time can be zero, if two dimensions are not interacting; but, the apparent time is not zero since the interaction of some dimensions will be occurring even when two particular dimensions might not be interacting. Since time has subsets, it is not a dimension by itself, no more than that of matter (which is three dimensional) be referred to as one dimensional, namely having only the dimension of space.

Chapter One: The Theory of Dimensionality:

A Unified Theory of Dimensions, Space, Strings, Matter,

Energy, Light, and Time

Part 4: The Cosmic Component

Black Holes: According to the theory of general relativity, at the core of a black hole is an extremely compact mass that deforms space-time. This, of course, is based upon the assumption that time is a dimension in and of itself (used to be referred to as the fourth dimension). Consequently, since time is perceived as being one dimensional, the event horizon around the black hole presents a problem. If time is but one dimension, time appears to stand still at the event horizon even as the substance to be absorbed is moving towards the black hole. Yet, if time is the product of the interaction of dimensions, a different story emerges. Time appears to stand still in that many dimensional interactions are negated by the extreme force of gravity from the black hole. That does not mean that time stands still. In contrast, while interactions among some dimensions are eliminated, other interactions are, in fact, accelerated. As such, time does not stand still; it only appears to do so in relation to what the observer is using as a benchmark for time. Even the very core of the black hole is still subject to time in that the black hole is moving in space and, thus, an element of interacting dimensions still exists within a black hole. Yet, within a black hole, the number of dimensional interactions decreases and approaches the 0 dimension; but, since the black hole is moving in space, time in the black hole never reaches the 0 dimension.

When Galaxies Collide: The prevalent approach to the collision of galaxies is to focus upon the stars and their respective systems. The consequence of such a collision is theorized to be a super galaxy, much as will happen when the Milky Way galaxy and the Andromeda galaxy collide in the distant future. But, galaxies are more than collections of stars, solar systems, and other matter. Galaxies also contain numerous black holes; one in particular, at the center of each galaxy, is of super-sized proportions. The presence of the black holes, and in particular the gigantic one in the centers of each galaxy, has a significant impact on the dimensions and time when galaxies do collide. When the galaxies collide, the black holes will encounter new stars, energy and matter upon which to exert their tremendous gravitational pulls. On one level, one can visualize millions of black holes of each galaxy consuming parts of the corresponding galaxy. The enormous black holes at the centers of each galaxy will have a virtual cosmic feast as new material falls within their gravitational pulls. But, it is not only their impact on energy and matter that is important; the black holes also interact with each other. Eventually, even the super-sized black holes at the centers of each galaxy will exert gravitational pull on each other. A moment occurs when the event horizons interact. This is due to the collision of the black holes. When that happens, matter and light are trapped between two black holes and may be unable to be forced to go to either black hole. It is when that happens that, from the standpoint of the object caught between the black holes, that the interaction of the dimensions comes closest to zero and, as such, time appears to stand still. Of course, time never quite stands still, since the galaxies, with their respective black holes and other contents, are still moving within the space of the universe.

Going Backward in Time: Time is a measure of the interactions of dimensions. There are two ways in which something can appear to go backward in time. The first way is by the using energy to do so; in essence, energy and/ or matter would have to be destroyed in order for something to go into the past. The ultimate destination would be the return to the origin of creation whereupon no matter, dimensional interactions, or time would exist. If something existed in the 0 dimension, it would, of course, be outside of the interaction of dimensions. In the 0 dimension, there is no real time, either forward or backward. If entry into the 0 dimension were possible, one could conceivably go forward or backward at will. But then, one would not have any matter to do so. The second way is by apparently going against (or backward in) time. That would occur if something were firmly anchored in two or more dimensions, whilst everything else around it was not anchored in such dimensions. As such, it would still be going forward in time, but at a different rate than everything around it. From the viewpoint of everything around the fixed object, it would appear that the object would be going reverse in time. But, in fact, it would still be going forward in time. The two means can be described by using the analogy of a stream with the water in the stream going forward in time. The first means of going backward in time would be like fish swimming upstream. From the standpoint of the water, the fish would be going back in time. And, to do so, the fish would be expending energy to do so. The second means would be like a rock solidly stationed in that stream. From the standpoint of the water in the stream, the rock would appear to being going backward in time. But, it is not going upstream, it is merely stationary. The appearance of the rock going backward in time would be only an illusion of going backward in time.

Chapter One: The Theory of Dimensionality:

A Unified Theory of Dimensions, Space, Strings, Matter,

Energy, Light, and Time

Part 5: Dark Matter and Dark Energy

Copyrights © 2011, 2019, 2020, James L. Jordan, PhD, PhD

and Deovina N. Jordan, PhD, MD

Dark Matter: Consists of components of light matter at absolute zero (at which atomic motion - e.g., that of electrons and inter-nuclear particles - ceases to exist).

Can arise from:
1.   Components of light matter being formed, but not energized yet (still undergoing reactions). As such, dark matter would be the precursor to light matter. Since dark matter is twice

as abundant as light matter, the implication could be that the universe is still young.

2.   Components of light matter which no longer undergo reactions from dispersal of energy into the surroundings.   As such, dark matter would be the end of light matter.   In that dark matter is twice as abundant as light matter, the implication could be that the universe is old and possibly dying.

3.   Both 1 and 2 above.   In such case, little can be concluded about whether the universe is young or if it is old and possibly dying.

In terms of time, dark matter has virtually no time component in itself (many of the dimensions are no longer interacting, particularly at the atomic and subatomic levels).   A very small time component exists in that it is part of a larger environment (e.g., galaxy) which moves relative to its surroundings (thus giving rise to a dimensional interaction which imparts a time component even to dark matter).

In terms of anti-matter, dark matter is not anti-matter.   It may be a precursor to anti-matter, depending upon what triggers the interactions (e.g., in positive dimensions or in negative [anti-] dimensions).

If a photon was to encounter dark matter, it would be by itself, lacking sufficient matter to cause movement in the hole of the dark matter. The energy of the photon would dissipate and the dark matter would remain as it was before. Because of that, dark matter would not exist directly adjacent to a star. In that case, sufficient energy would be expended to cause particle and atomic motion. Then the interaction of dimensions would increase and time would subsequently accelerate.

An Expanding Universe and the Expansion of Dark Matter: For the universe to expand, notably at an increasing velocity (acceleration), energy must be expended to do so. Since energy cannot be created, it must come from somewhere. Moreover, that energy is being expended in an ever increasing volume (space). So, how can this be? Matter is being converted into energy (decrystalized) and/ or losing its energy (eventually becoming dark matter). Either or both events could be occurring and would propel the universe into an ever increasing volume of space. What is the net result on time? In one sense, time would accelerate; namely, there is increased interaction of the dimensions in terms of space overall (expanding universe) and at the subatomic level (conversion of matter into energy). But, other forces may lead to a deceleration of time, namely the conversion of light matter into dark matter, in which the interaction among dimensions decreases.

The Rise of Dark Energy: The existence of both positive and negative dimensions concurrently gives rise to the development of positive energy from the positive dimensions and negative energy (or dark energy) from the negative dimensions. In terms of time, since time is the result of interacting dimensions, decreasing or increasing in proportion to the level (degree, amount) of interactions among those dimensions, time is not dependent upon the negative or positive status of the dimensions involved. But, energy (positive or negative [dark]) and matter (positive or anti-matter), in contrast to time, are dependent upon the negative or

30

positive status of the dimensions. But, it should always be realized that positive matter or energy only exists in the perspective of the observer and in accordance with its opposite. This means that something consisting of anti-matter or dark energy would behave as if matter and positive energy were in fact anti-matter or dark energy.

Chapter One: The Theory of Dimensionality:

A Unified Theory of Dimensions, Space, Strings, Matter,

Energy, Light, and Time

Part 6: Light

Copyrights © 2011, 2019, 2020, James L. Jordan, PhD, PhD

and Deovina N. Jordan, PhD, MD

All Rights Reserved.

The Nature of Light/ Electromagnetic Radiation: Light has been known to possess properties of both particulate and wave nature. The wave nature is expressed in different wavelengths throughout the electromagnetic spectrum. Since the speed of light is constant, the wavelength feature cannot depend upon the speed of light. The wavelength of light is formed when a photon, traveling at the speed of light, spins in space. The spinning creates a de facto wavelength, particularly if the photon is asymmetrical (in space and/ or mass). The actual length of the wavelength is dependent upon the rate of spin as the photon travels through space. A slower spin results in a longer wavelength; a faster spin results in a shorter wavelength. When a photon encounters matter, it can impart some of that spin to the matter; that would result in light/ electromagnetic radiation with a longer wavelength. If the photon encounters matter, but obtains energy from that matter, that would result in light/ electromagnetic radiation with a shorter wavelength. If no energy is transferred, the wavelength remains the same. When two beams of light, of the same wavelength, are in direct opposition to each other, they can cancel each other out. This results from the photons colliding and having their spins negated; thus, their wavelength nature ceases. The photon then disintegrates.

What are the dimensional properties of light? Light has the properties of a photon which is three dimensional. It travels in a direction; that is 1st dimension. It spins; that is 2nd dimension. Consequently, light is actually 6th dimensional when it is traveling in a vacuum without any outside influence (e.g., black hole). If its path is distorted by a strong gravitational pull, such as from a black hole, light can become 7th or 8th dimensional, depending upon whether its change in direction is planar or spatial. Moreover, in terms of traveling through space, the strings with neutral point of interaction (POI) status would not offer resistance to the movement of the photon, either in tumbling or in trajectory fashion. As such, light can travel at the same speed without change if it is not affected by gravity and if it does not interact with matter.

The concept that light is faster than anything else applies

only to matter. At speeds faster than the speed of light, matter disintegrates into energy, but that does not mean that energy is not capable of going faster than the speed of light. It only means that matter cannot go faster than that and still remain matter. Matter can be thought of as being decelerated energy.

Variance in the interaction of dimensions adds an important component to light; namely, light does not always travel the same distance in a given length of time. It also impacts such equations as $e = mc^2$; that equation is dependent upon the assumption of time as being constant. But, what if time changes? First, the distance traveled per second (which would be a standard on earth) would change. The earth standard, however, is still usable as a measure to quantify light, energy, and distance; but, that does not make it the norm throughout the universe. The importance of the earth standard is that on earth, gravitational forces are constant, input of energy from the sun is basically constant, energy on the earth is constant, the space is constant, the amount of matter is constant, and, as such, observed time appears constant. Note that the constancy observed in space is also that which occurs for the solar system. This permits the development and application of a number of equations, such as $e = mc^2$ or $ek = \frac{1}{2} mv^2$; their development and application depend upon a constant observed time which, fortunately, is the current case on planet earth and the solar system. It will, however, change when the sun dies.

Differences in Distances Traveled by Light: Assume that one was an observer watching two beams of light. Originally, the two beams of light would travel the same distance per unit of time observation. But, let us assume that one beam of light was to enter a spatial realm in which the interactions of the dimensions were less; in which case, the time to which that beam of light would be exposed will be slower than the beam going through the original interactions of dimensions (in which time remained constant). The apparent distance traveled would appear to be less per unit of time (as per the observer) for the beam of light in the realm with lesser dimensional interactions than that with the unchanged dimensional interactions. In actuality, the beam of light would still be traveling at the same speed, distance per unit of time, but the units of time would change according to the interactions of the dimensions in which it is traveling. The converse is true if one beam of light were to enter a realm in which the interactions were greater; in which case, the observer would assume the beam which entered the realm with greater interactions of dimensions would be traveling faster. But, in fact, both beams of light would still be traveling at the same speed; just the units of time would change according to the interactions of the dimensions.

The Photonic spin: When viewed as sine curves, with the intersection at the x-axis (the direction in which the photons would be traveling), photons can have either a positive spin or a negative spin. The spin has major consequences on how the photons interact. Two photons, of the same sine curve and with the same wavelength (and amplitude), would cancel each other out if they were to directly collide with each other. This is because the spins of the photon would be similar to that of seeking to turn a bolt equally in two opposite forces of the same magnitude. But, what if the two photons, of opposite sine curves and with the same wavelength (and amplitude), were to interact? The amplitudes and the wavelengths would increase the wavelengths for both photons.

36

Currently, theory holds that a photonic spin accounts for the polarity of light. But, the same theory did not account for the possibility of the asymmetry of the photon. Neither did prior theory account for the possibility of multiple spins. As noted earlier, the spin of an asymmetric item, with the axis corresponding to the greatest length of that item, would result in a wavelength pattern as it moved forward in a line (vector). Differences in direction of the spin can be held to account for polarity if that one direction of spin would be negative and the other positive.

Directions of two spins are theoretically possible for the photon. One spin would be according to the greatest length/ depth of the particle (e.g., photon) and the other according to the smallest length/ depth of that particle. The first spin has its direction in accordance to the greatest length (affecting wavelength and amplitude [as explained later in this book]). Two different, opposite, directions are possible. This contributes to the development of polarity which forms according to the direction of the two spins (one called positive and the other negative). This spin can be visualized as a man doing cartwheels. He can do a cartwheel to one direction or the other, but not both at the same time. The other spin can also occur perpendicular to that, similar to an ice skater performing a spin. Again, two directions are possible for that ice skater; likewise, two directions would be possible for the photon. It is also possible that both spins, completely perpendicular to each other can occur simultaneously. That would result in four possible spin combinations for a photon.

The Photonic Molecule? One approach to the photon is to visualize it as being a collection of particles, basically a molecule of sorts (though not by atoms). The problem is that this would lead to great variances in the size of photons. While this may explain the energy differences between photons, it does not explain the dual nature of light, namely particle and wave. A photonic molecule has another problem. If asymmetry occurs both in length and width of the particle (such as asymmetry occurring in a man's bodily dimensions of height and girth), then a double spin can occur, resulting in a double wavelength. One wavelength would be the primary wavelength which would conceal the secondary (shorter wavelength). But, dual wavelength light has yet to be detected.

The simpler approach is to visualize a photon as being a particle that is asymmetrical spinning in space going in a particular direction at a defined speed. Another simpler approach is to visualize a photon that is comprised of two sub-particles, one of which is larger than the other. The spin would still create the wave, while the particulate nature of light would remain. If the latter is true, then one should be able to split the photon. But, this has not happened, leading support to a single asymmetrical particle.

How About a Photon That Stops Traveling? If a photon is stopped, and no longer travels at the "speed of light", is it still light? No, it becomes a particle. Light, dark or non-dark, exists only when the photon is spinning and traveling at the "speed of light."

Dark light: Could dark light exist? Definitely. If light is an asymmetric particle spinning and traveling basically in a straight line in space at a particular speed, then a mirror image of that particle can exist. Mirror images can arise whenever asymmetricity exists in three dimensions. Simply turning a photon in a two-dimensional context would not result in a new asymmetry no more than turning this page upside down would result in a new asymmetry.

Asymmetry can occur by one or both of two ways. First, the photon can be asymmetric in space. For example, a bowling pin would represent such an asymmetry. When the bowling pin would spin in space, the larger end (the base) would be that which would appear to be making waves as it moved through space. Second, the photon can be asymmetric in charge. Even a sphere can be so asymmetric if one side of the sphere is positive and the other side is negative. It would be much like a planet (not a true sphere, of course) had the north pole positively charged and the south pole negatively charged. If the photon was asymmetric, much like the bowling pin, and the ends charged, then dark light would be the result of a change in polarity of the photon from that held by non-dark light.

Would dark light have a different effect on the interaction of dimensions, and thus on time, than non-dark light? Yes, it would if it interacted with non-dark light so that both canceled themselves out. But, in the absence of interaction with matter or non-dark light, it would interact with the dimensions in the same manner as non-dark light. However, if the polarities of the charges on a photon are different between dark and non-dark light, the interaction with matter (either matter or anti-matter) would also change. How different the interaction would be cannot be ascertained since dark light itself has not yet been measured or studied. Note, that throughout this theory, unless light is specifically referred to as being dark light, it is non-dark light, the light we normally encounter and call light.

Chapter One: The Theory of Dimensionality:

A Unified Theory of Dimensions, Space, Strings, Matter,

Energy, Light, and Time

Part 7: Quantum Physics, Particles and Time

Quantum Physics and Particles in Space: The dilemma of a single particle in two or more places at the same moment of time occurs with quantum mechanics (physics); this occurs because of the alignment of strings, corresponding to a particle's position in space. The strings do not always realign themselves instantaneously after a particle leaves and changes position. Moreover, it may result from the interaction with other forces; particles do not act independently of their environment.

How can a particle be in more than one place at the same time? The answer lies in the overlapping trajectories of a particle as it travels/ interacts with different dimensions. When a particle goes through space, around another object such as the electron around a nucleus, its movement through space directly influences the strings corresponding to its location. The strings, however, do not reorient themselves immediately to the previous position (of the particle's transit) after the passing of that particle. This effect is compounded when the particle's trajectory directly influences the same region of strings which had been previously stimulated, but which have not returned completely to the original, non-stimulated state. Each time this happens (from overlapping, dissecting, or near-proximity trajectory intersections), the effect becomes more pronounced. In time, the actual location of the particle becomes confusing in that its signature will be noted at a number of places in space. This gives the image of a particle being in more than one place, which may be quite distant to each other (in terms of particles and paths taken), at the same instance.

Why does time travel in a positive direction? Time is the degree to which dimensions interact with each other. If the dimensions were to reverse in their interaction with each other, the result would still be an interaction of the dimensions. The result is that time would still travel in a positive direction. Note: The directions of the interactions of the dimensions (whether forward or backward, interaction or reversal of interaction) are of no consequence; time itself would be an absolute value and have a positive value in either direction of dimensional interactivity. Let us assume a person wanted to go back (and could go back) to the moment the universe was created, the big bang, and could unravel all dimensional interactions that had occurred up to the point of reversal. It would be understood that from the point of the big bang to the reversal, time would be positive. But, from the point of the reversal back to the big bang, the dimensions, even though reversing in their interaction from before, would still be interacting, and, as such, time would still proceed in a positive direction.

Conclusion: Time is not the fourth dimension, but arises from the interaction of dimensions. To date, time has not really been defined, but assumed. By defining time, it becomes possible to better understand physics and other disciplines which incorporate time into their equations and reasoning. Since the interaction of dimensions appears to be relatively constant to a present observer on earth, time becomes a relative constant and, as such, no physics or other equations incorporating time as a constant are voided or transformed (at least from an earthly and current perspective).

Chapter Two: Particle Spin and Wave Formation

Jordans' First Law of Particle Spin and Wave Formation: A particle spinning in space, traveling along a vector, develops an apparent wave (henceforth called "wave"). The amplitude of the wave formed is dependent upon the length of the particle perpendicular to the vector upon which it travels. As such, two particles, rotating or spinning at the same speed and traveling at the same velocity along a vector of travel, will create the same wavelength even though one may be much larger/ longer than the other. The reason is that the wave for both has the highest amplitude away from the vector of travel at exactly the same time. There is a difference in the intersection locations of the two waves with the vector of travel, but that phase shift is constant and does not impact the actual wavelength. As such, every circle or sphere with the same speed of rotation traveling at the same velocity along the vector of travel creates the same wavelength as every other circle or sphere with the same speed of rotation traveling at the same velocity along the vector of travel. From the standpoint of light, velocity along the vector of travel is 299,792,459 m/s. As such, given a certain distance, such as the speed of light at about 9.4 trillion kilometers, two photons, rotating at the same speed, but having different sizes, would present with the same wavelength.

Another way to picture this phenomenon would be to picture a rolling car wheel. Put a mark on the outermost portion of the wheel, another three-fourths away from the center, another at half way to the center, and another at a fourth of the distance from the center. The illustration becomes easiest if those points are in a straight line in relation to each other. In one rotation of that wheel, each of the points would be in the same relative position to the center of the tire throughout that rotation. While the wheel moved a certain direction, each point maintained the same position regarding its location. If those points were to be drawn onto a plane during the rotation, a wave pattern would be presented for each of the points (except the exact center) as it moved with the rotating tire. And, each wave would have exactly the same wavelength. Even though the wavelength would be the same, the amplitudes of the waves would differ considerably. If the object rotating was a sphere (such as a ball) or a cylinder, every point on or in that sphere or cylinder, (except the exact center or the line that is precisely perpendicular to the vector of travel) would create the same wavelength pattern as every other point on or in that sphere or cylinder as the sphere moves along at the same velocity along a vector of travel. As such, consistency in wave propagation, in terms of wavelength, would occur regardless of the size of the sphere or cylinder. Note, however, that the amplitudes of the produced waves would vary considerably among different sized spheres or points at different lengths from the center in a sphere or cylinder.

Differing wavelengths are created by particles rotating or spinning at different speeds while traveling the same distance on a vector path. As such, if a star rotated at a thousand times per second and a proton also rotated at a thousand times per second, both would have the same wavelength over the same distance during the same period of time of vector travel. The amplitudes of their waves would be immensely different, but the wavelengths that they have would be exactly the same. Of course, the wavelength of the particle would appear to be virtually flat, whereas that of the star would exhibit great fluctuations in deviation from the vector of travel. This is the reason why light can present properties of both matter and wave motion. The amplitude of the light wave is determined by the length of the photon perpendicular to the vector of travel. The wavelength of the light wave is determined by how fast the photon is spinning or rotating while traveling a specific distance on that vector of travel. That is also why an electron has a wavelength a hundred times shorter than a photon. The photon is spinning or rotating a hundred times slower than the electron. As such, when the photon and the electron travel the same distance on the vector of travel, the photon develops a wavelength that is a hundred times longer than that developed by the electron.

The spin or rotation of a photon is directly related to the energy contained in that photon. As such, if a photon is spinning or rotating faster, it has more energy than a photon which is spinning or rotating slower, even though they are traveling at the same speed of light along the vector of travel. If the photon interacts with another photon, it can gain speed of spin or rotation by making the other photon slower. Or, it can be made slower by contributing to the spin or rotational speed of the other photon. Likewise, the spin or rotation of a photon can be sped up or slowed down in accordance with its interaction with chemicals which can impart energy to or take energy from the spinning or rotating photon. Note that the photon does not change velocity in the vector of travel, only its spin or rotation.

If a particle spins on only one end of the line/ plane perpendicular to the vector of travel, it presents with only one wavelength. Only one wavelength is formed by the spinning or rotating particle even if multiple waves are being generated (due to differences from the center of spin or rotation). The reason is that all points along that line/ plane are actually spinning or rotating at the same speed in regard to the vector of travel. The amplitude of the waves generated do differ. Yet, one sees only one wavelength even when multiple waves are being generated by the same spinning or rotating particle.

Jordans' Second Law of Particle Spin and Wave Formation: Closely related to the formation of waves by particles with different rates of spin or rotation at the same vector of travel velocity is what happens when the vector of travel velocity changes for particles with the same rate of spin or rotation. Of course, since the vector of travel velocity would no longer be at the speed of light, the energy would no longer be electromagnetic, but wave formation is still affected. While spin or rotation determines the wavelength of a particle traveling at the speed of

49

light in its vector of travel, the wave frequency, of two particles with the same rate of spin or rotation, changes when the vector of travel velocity changes over a specified distance. This only applies when the particle spin is the determinant factor in the vector of travel velocity. The reason is that the same number of wavelengths is required to cover a given distance. All points on the particle generating a wavelength travel the same distance per unit of time.

To illustrate, this can be observed when all points on a car tire reach the same destination (a given distance away) with the same number of car tire rotations (corresponding to waves). Simply put, a vehicle traveling from point A to point B gets to point B later when traveling at a slower velocity than a vehicle traveling at a faster velocity. It does so because the particle (the wheel in this case) is turning slower for the vehicle that arrives at point B later. In essence, the particle spin or rotation determines the particle's velocity along the vector of travel.

Jordans' Third Law of Particle Spin and Wave Formation: In the final case, the spin or rotation of two particles remains the same, while the vector of travel velocity changes; the distance covered is determined by time, not by a predetermined travel distance along the vector of travel. This is because for a given vector of travel velocity, the particles spinning in a given unit of time cover different distances.

In such a scenario, the wavelength is stretched out/ elongated in the particle traveling at a faster rate of particle spin. In contrast, the wavelength is compressed in the particle traveling at a slower rate of particle spin. A longer wavelength simply covers more distance per spin than a shorter wavelength. The distance covered compounds with the number of spins. At the same period of time, the longer wavelength would cover more distance. This would be like a person traveling at one rate, walking on the ground or walking while inside an airplane. For every step (a wavelength analogy, though not precise), he would cover less distance while walking by himself than while walking in a moving airplane. Basically, for **Jordans' Third Law of Particle Spin and Wave Formation**, an interaction of dimensions is occurring. The man walking in the plane (spin of the particle) involves three dimensions, his movement in the plane involves three more dimensions (body moving in space), and the plane moving in space involves another three dimensions. And, since time represents an interaction of dimensions, the time component is the most complicated interaction for the third law of particle spin and wave formation (in contrast to the first two laws of particle spin and wave formation). The particle spinning involves three dimensions. Its movement in space involves three dimensions. And when the vector of travel velocity changes, other dimensions are also being affected (two to three dimensions).

Another useful analogy to describe what is happening is that of two bullets fired from two different rifles. Both bullets can have the same spin when leaving the rifle muzzle, but one bullet may be propelled faster than the other bullet. The fact that the bullets are being propelled does not change the bullet spins (that results from the rifling in the barrel). The spins by themselves would produce the same wavelength if the velocity of the vector of travel was the same. But, the velocity of vector of travel has now changed. As such, for every spin of the faster bullet, more distance would be covered than for the slower bullet. Since wavelength has built in it the rotation per distance, the faster bullet would go further per spin than the slower bullet. The wavelength formed by the actual spin is one component and the change in the vector of travel velocity is another component in the overall wavelength formed by the particle.

In space, there is no medium upon which a particle spin would determine velocity along the direction of the vector of travel. But, the vector of travel can change. For example, light is affected by the gravitational pull of a black hole; light becomes curved. This causes two problems. Nothing goes faster than the speed of light and a curve involves acceleration. The net result, to conform to the two caveats just mentioned, is that time itself changes along with a change in the overall wavelength of particle spin. The black hole, in essence, is acting like the plane in which the man (the particle) is walking in (the spin). This is still in accord with the *Jordans' First Law of Particle Spin and Wave Formation*, but the overall wavelength becomes stretched out with the influence of the black hole. Because light in a black hole becomes curved and a curve involves acceleration, the velocity of the particle, traveling along the vector of travel, changes.

Conclusion: *Jordans' Three Laws of Particle Spin and Wave Formation* are as follows:

- The first law, *Jordans' First Law of Particle Spin and Wave Formation*, states that wavelength is determined by the rate of spin of a particle traveling at a specific or constant vector of travel velocity (e.g., the speed of light).

- The second law, ***Jordans' Second Law of Particle Spin and Wave Formation***, states that wave frequency is determined by a particle spinning at a constant rate but at a different vector of travel velocity over a specified distance. This only applies when the particle spin is the determinant factor in the vector of travel velocity.

- The third law, ***Jordans' Third Law of Particle Spin and Wave Formation***, states that overall wavelength is affected by a particle spinning with a constant wavelength but moving in a different vector of travel velocity.

Chapter 3: Black Holes, Quasars, and the Origin and Fate of the Universe

Copyrights © 2019, 2020, James L. Jordan, PhD, PhD and Deovina N. Jordan, PhD, MD

Particles of matter, both light and dark matter, when released into a zero-gravity environment, clump together. That tendency of matter to aggregate has led to the formation of meteorites, asteroids, planetoids, planets, suns, solar systems, and even galaxies. Black holes are simply a result of that aggregation, carried to an extreme. The nature of the black hole is not a function of its size (volume), but of what went into its formation. To exert the same level of gravitational forces, a black hole formed from a massive aggregation of light elements (e.g., hydrogen, helium) would be larger in comparison to one composed from the accumulation of heavy elements (e.g., iron and heavier elements). That which has greater mass exerts a greater gravitational pull than that which has lesser mass. The earth exerts a greater gravitational pull than the moon. The sun exerts a greater gravitational pull than the earth. The same holds true for that matter/ mass which is pulled into a black hole. The more the gravitational pull of that matter/ mass which is used in the formation of the black hole, the greater the gravitational pull by that black hole in proportion to its size.

Black holes, at the center of galaxies, are known to bring in not only light, but also whole stars and solar systems. Theoretically, such could create a singularity in which everything becomes compressed to a degree that is infinitesimally dense and small. But, such does not occur. When the matter is being compressed into the black hole, it continues in a swirling fashion, just as the galaxy around it is swirling. In fact, the directions of swirls inside and outside the black hole are the same. A result of the swirling is the creation of poles due to magnetism. As such, the energy within a black hole becomes directionalized. When a sufficient amount of energy and matter are so concentrated and directionalized, they escape the black hole via a quasar. The quasar is composed of plasma that had been constructed from matter of differing atomic elements. As such, iron, known for being responsible for the death of stars, can then be reconverted back to hydrogen. When the reconversion has occurred, then new stars can develop from the newly formed hydrogen. In essence, therefore, a black hole functions as a recycling center for the galaxy and the quasar is the actual recycling process in action.

The quasar is directional. It is in a 90 degree orientation (perpendicular) to the planar spin of the galaxy. As a consequence, matter and energy can feed from the galaxy into the black hole without being destroyed by the quasar. From an earthly perspective, no quasar generated by the Milky Way galaxy threatens earth's existence. Earth is at the periphery of the galaxy spin and in line with its planar orientation. A quasar from the Milky Way galaxy's black hole would be oriented away from the earth, not toward it. Second, when a sufficient amount of plasma has been ejected and matter reformation has occurred, the galaxy can be reconstituted. Such is possible since the quasar does not extend out into infinity. Instead, it dissipates into a massive gas cloud that is at right angles to the original galaxy. The evidence for that exists in the gaseous clouds of the Milky Way galaxy that lie perpendicular to the rest of the galaxy and in the orientation of a past quasar. A third consequence is that a black hole may eject a quasar for a period of time and be quiescent during other periods of time. The black hole expands during the collection phase and shrinks during the quasar phase. But, since the quasar dissipates into a gas cloud, a quasar from a distant galaxy is of no threat to another galaxy. Thus, earth is not in any danger from a quasar that originates from the black hole of another galaxy.

This process has notable consequences for the universe itself. One theory is that an accumulation of iron in stars will eventually cause so many stars to die that the universe itself will perish from the continual loss of active stars. But, with the operation of black holes with their quasars, such an accumulation of iron and heavier elements does not occur. Instead, the iron and heavier elements are converted back into plasma and ejected back into the universe as plasma, whereupon the lighter elements, in particular hydrogen, are formed. Consequently, stars and galaxies can reform and the continuance of the universe is assured.

That black holes are involved in the continuance of the universe also brings upon another important issue. A black hole, not just a single particle of matter or component of energy, was involved in the beginning of the universe. Dark matter compressed in a black hole was responsible for the release of both the energy and the matter seen in the visible universe. The dark matter was transformed into non-dark matter and the associated energy. A single particle cannot reproduce itself. A single particle has never been observed exploding into a vast conglomeration of energy and matter. In fact, doing so would violate a law of thermodynamics that energy is neither created nor destroyed. If a single particle could so dramatically reproduce in the beginning of the universe, there would still be evidence of it now. But, particles are not undergoing big bangs, even though the number of particles has so greatly increased since the beginning of the universe. In contrast, black holes do release matter and energy. That is still being observed. In fact, the nature of quasars, as discussed earlier, supports such a hypothesis. As such, the likely origin of the visible universe is better explained by black holes rather than by an anomaly experienced with a single particle at a singular point in time.

As theorized in this book, black holes are the focal point of the universe and of physics. They are responsible for the origin of the universe. And, they are necessary for its continued existence.

Chapter Four: The Energy-Matter Cycle

Before there was light, there was dark matter which is even more dissociated and disorganized than plasma (a form of matter). But, since dark matter consists of the sub-particles responsible for forming the particles associated with plasma, it is also subject to the forces of gravity and other forces of physics akin to plasma. Dark matter consequently aggregated to form the first primordial black holes. From those primordial black holes, both light and light (non-dark) matter emerged. Note, that the original black holes are being called primordial since they consisted only of dark matter. Current black holes contain dark matter and plasma. Since black holes, both primordial and current, occupy space and are entities possessing dimensional characteristics (defined in chapter one), time eventually came into existence after the formation of the primordial black holes. That is because time is a product of the ratio of energy and matter (as explained in *Jordans' Law of Time*). Such an interaction evolved in the primordial black holes because energy interacted with the plasma that was formed in the primordial black holes. And, since the dimensions interact, the primordial black holes also interacted. When light and matter finally emerged, such was not a singular event involving a single particle. Instead, it was an event that involved many primordial black holes simultaneously ejecting (via jets) both light and matter (in the form of plasma which had accumulated in the original black holes). The plasma then formed the basis for the formation of all associated/ organized light (non-dark) matter (gas, liquid, solid).

The gravity of modern (non-primordial) black holes is known to draw in everything from stars to light into the black hole. Therein, the matter and energy become not only trapped but transformed. The light matter either becomes plasma or dark matter. Plasma or dark matter is important. There is a relationship between the two within a black hole. Dark matter can be converted into plasma or forms of energy (e.g., electromagnetic radiation/ light). Also, plasma or forms of energy (e.g., light) can be converted into dark matter. Note: Light itself is a particle that not only has velocity but also wavelength (formed from particle spin velocity as indicated in chapter two). The direction of the transformation (equation) is dependent on the concentration and size (therefore, gravitational forces) of dark matter versus plasma within the black hole. As a consequence, a series of events occurs within a black hole. Matter (dark and light) enters the black hole. Matter is dissociated into plasma. Plasma can be converted into dark matter or remain as plasma. The dark matter can also be converted into plasma. Those

61

conversions occur within the framework of extremely intense/ high gravity. When dark matter is above a threshold, the formation of plasma is favored. When the plasma reaches a critical point, in relation to the dark matter, the plasma is ejected (as a jet) from the black hole. That ejection is perpendicular to the accretion disc (where matter and light enter the black hole). By being perpendicular to the material entering the black hole, the path of least resistance occurs. As such, a black hole can be dormant, accumulating matter and light energy, regarding the jetting forth of plasma. But, it can become active, thereby forming quasars by which plasma is propelled outside of the black hole. Some of this was discussed earlier in chapter three.

Plasma is not the only thing that can be formed within a black hole or ejected from it. Anti-plasma can also be formed, depending upon the formation of particles within the black hole. If the particles are opposite of the particles necessary for the formation of plasma, the result is anti-plasma. Anti-plasma then can form anti-matter. Note: Anti-matter enters the black hole in the same manner (e.g., in response to gravity) as light matter. Moreover, when anti-matter and light matter interact, dark matter is created. As such, the reaction of anti-matter and light matter is not an extinguishing event of light matter or anti-matter. Instead, it represents the conversion of both into a truly unassociated/ dissociated form, namely dark matter. The great amount of energy released is the energy produced from the breakdown of structure for both the light matter and the anti-matter.

As noted at the beginning of this chapter, the primordial black holes eventually took on the form of modern black holes. This contributed to the formation of an energy-matter cycle

(*Jordans' Energy-Matter Cycle*) currently occurring in the universe. Starting with the black hole, the energy-matter cycle (*Jordans' Energy-Matter Cycle*) proceeds as follows:

1. Plasma is ejected from the black hole.

2. Plasma ejected from the black hole coalesces to form very light elements (primarily hydrogen).

3. The hydrogen coalesces to form stars. Because of the heat and gravity within a star, plasma exists at the core of a star. Note: Elements aggregate together when they are in a vacuum environment.

4. The star may be sucked into a black hole, causing a return of its components to plasma and dark matter. A return to step 1 eventually occurs.

5. The hydrogen of a star undergoes fusion to form heavier elements. Accumulation of iron and heavier elements leads to a star's death.

6. The dying star goes nova or forms a white star. Thus, even heavier elements are formed.

7. The material formed from a dying star can aggregate to form planets or other astronomical entities (e.g., asteroids). The material from a dying star can also be sucked into a black hole, causing it to convert to plasma and dark matter. A return to step 1 eventually occurs.

As indicated in chapter three, the black hole actually functions as a cosmic recycling center. The universe does not die out or simply fade into nonexistence (as some have claimed because of so many stars dying). Instead, the components of the universe are replenished due to the activity within black holes. Galaxies are created and are even reoriented (according to the orientation of the plasma relative to the rest of the galaxy).